SandCastle™
Sports Gear

HELMETS, MASKS & GOGGLES

MARY ELIZABETH SALZMANN

Consulting Editor, Diane Craig, M.A./Reading Specialist

A Division of ABDO

ABDO
Publishing Company

visit us at www.abdopublishing.com

Published by ABDO Publishing Company, a division of ABDO, P.O. Box 398166, Minneapolis, Minnesota 55439. Copyright © 2012 by Abdo Consulting Group, Inc. International copyrights reserved in all countries. No part of this book may be reproduced in any form without written permission from the publisher. SandCastle™ is a trademark and logo of ABDO Publishing Company.

Printed in the United States of America, North Mankato, Minnesota
062011
092011

 PRINTED ON RECYCLED PAPER

Editor: Katherine Hengel
Content Developer: Nancy Tuminelly
Design and Production: Anders Hanson
Image research: Stacy Nesbitt
Photo Credits: Thinkstock (Jupiter Images, Zedcor Wholly Owned, Darrin Klimek), Shutterstock

Library of Congress Cataloging-in-Publication Data
Salzmann, Mary Elizabeth, 1968-
 Helmets, masks & goggles / Mary Elizabeth Salzmann.
    p. cm. -- (Sports gear)
 ISBN 978-1-61714-826-2
 1. Sporting goods--Juvenile literature. I. Title. II. Title: Helmets, masks and goggles.
 GV745.S36 2012
 688.7´6--dc22
                        2010053289

SANDCASTLE™ LEVEL: TRANSITIONAL

SandCastle™ books are created by a team of professional educators, reading specialists, and content developers around five essential components—phonemic awareness, phonics, vocabulary, text comprehension, and fluency—to assist young readers as they develop reading skills and strategies and increase their general knowledge. All books are written, reviewed, and leveled for guided reading, early reading intervention, and Accelerated Reader® programs for use in shared, guided, and independent reading and writing activities to support a balanced approach to literacy instruction. The SandCastle™ series has four levels that correspond to early literacy development. The levels are provided to help teachers and parents select appropriate books for young readers.

Emerging Readers
(no flags)

Beginning Readers
(1 flag)

Transitional Readers
(2 flags)

Fluent Readers
(3 flags)

CONTENTS

HELMETS, MASKS & GOGGLES ?

Helmets, masks, and goggles
are sports gear.

Helmets **protect** heads. Masks protect faces. Goggles protect eyes.

BASEBALL HELMETS

Batters wear helmets.

The helmet
covers the head
and ears.

BATTING HELMET

Catchers wear helmets too.

CATCHER'S HELMET

FOOTBALL HELMET

Football helmets have face masks. They have chin straps too.

FACE MASK

CHIN STRAP

Some football helmets have **visors.**

VISOR

ICE HOCKEY HELMET

Ice hockey players wear helmets.

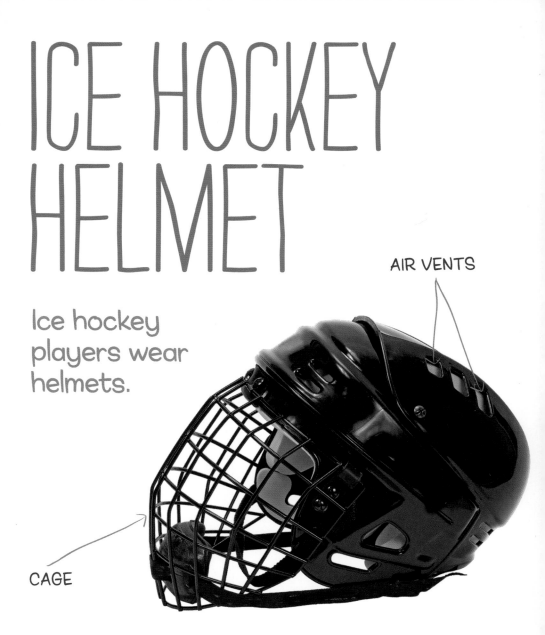

AIR VENTS

CAGE

Goalie helmets have **wire** masks.

RIDING HELMET

People who ride horses wear helmets.

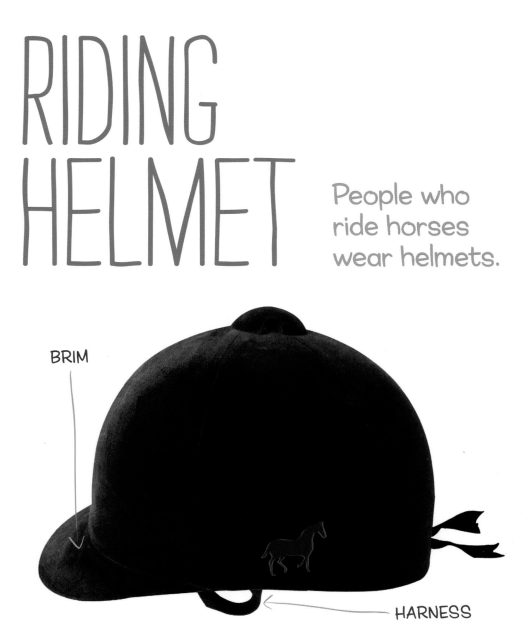

BRIM

HARNESS

Riding helmets have small **brims**.

SKI HELMET & GOGGLES

Helmets keep skiers safe and warm.

AIR VENTS

CHIN STRAP

Most skiers wear goggles.

BICYCLE HELMETS

Most bike helmets are light.

AIR VENTS

STRAPS

VISOR

BMX helmets are bigger.
They have masks.

SWIMMING GOGGLES

Swimmers wear goggles.

Goggles keep the water out!

RACQUETBALL EYE GUARDS

Eye guards have strong lenses.

STRAP

LENSES

Rackets and balls move quickly! Eye guards **protect** eyes.

FUN FACTS

- Old football helmets were made of leather.

- NHL players started wearing helmets in 1979.

- Some ski helmets have speakers. Skiers can listen to music!